discovermore

Space Science

Exploring the Milky Way

Elizabeth Pruet

Britannica
Educational Publishing
IN ASSOCIATION WITH
THE ROSEN PUBLISHING GROUP

Published in 2026 by Britannica Educational Publishing (a trademark of Encyclopædia Britannica, Inc.) in association with The Rosen Publishing Group, Inc.
2544 Clinton Street, Buffalo, NY 14224

Distributed exclusively by Rosen Publishing.
To see additional Britannica Educational Publishing titles, go to rosenpublishing.com.

Editor: Danielle Haynes
Book Design: Michael Flynn

Photo Credits: Cover Triff/Shutterstock.com; (series background) Dai Yim/Shutterstock.com; p. 4 Harleyzhang/Shutterstock.com; pp. 5, 10, 11, 17 (top), 20, 21, 23, 24, 25, 26, 27 courtesy fo NASA; p. 7 (top) Spacecraft: ESA/ATG medialab; p. 7 (bottom) R. Ibata (Strasbourg Observatory, ULP) et al., 2MASS, NASA; p. 8 Blue bee/Shutterstock.com; pp. 9, 22 VectorMine/Shutterstock.com; p. 12 BlueRingMedia/Shutterstock.com; p. 13 ESA/Herschel/PACS; p. 14 khlongwangchao/Shutterstock.com; p. 15 Mindaugas Gaspa/Shutterstock.com; p. 17 (bottom) Vadim Sadovski/Shutterstock.com; p. 18 Jemastock/Shutterstock.com; p. 19 https://commons.wikimedia.org/wiki/File:Kepler-452b_artist_concept.jpg; p. 28 KIDSADA PHOTO/Shutterstock.com; p. 29 AstroStar/Shutterstock.com.

Library of Congress Cataloging-in-Publication Data

Names: Pruet, Elizabeth author
Title: Exploring the Milky Way / Elizabeth Pruet.
Other titles: Discover more. Space science
Description: [Buffalo] : Britannica Educational Publishing, [2026] | Series: Discover more: space science | Includes bibliographical references and index.
Identifiers: LCCN 2025038964 | ISBN 9781641905275 library binding | ISBN 9781641905268 paperback | ISBN 9781641905282 ebook
Subjects: LCSH: Milky Way--Juvenile literature
Classification: LCC QB857.7 .P78 2026
LC record available at https://lccn.loc.gov/2025038964

Manufactured in the United States of America

Some of the images in this book illustrate individuals who are models. The depictions do not imply actual situations or events.

CPSIA Compliance Information: Batch #CWBRIT26. For further information contact Rosen Publishing at 1-800-237-9932.

Contents

Meet the Milky Way

If you ever find yourself out in the country in an area that's extra dark, take a look up in the night sky. See that faint dusting of a white "cloud" stretching across the sky? That's the Milky Way galaxy. It's made up of hundreds of billions of stars and clouds of dust and gas. Our sun is one of those billions of stars. That means Earth and all the other planets of our solar system are also part of the Milky Way.

It's easier to see the Milky Way in places where it's very dark. Big cities can be too bright with light pollution.

We don't have a full, true photograph of the Milky Way because a telescope hasn't traveled far enough away to capture it in its entirety. This is a drawing.

The galaxy is shaped like a giant pinwheel. We're inside one of the pinwheel's arms. All the stars you see with the naked eye are in our arm or elsewhere in the pinwheel. You can only see stars outside our galaxy using a telescope.

Consider This

The Milky Way galaxy is full of dust, as well as stars. With the naked eye, we can see only a few thousand of the galaxy's hundreds of billions of stars. What could be blocking our view?

Measuring Our Galaxy

If we can't see the full Milky Way from Earth, do we know how big it is? Yes! Distances in space are measured in light-years. One light-year is the distance that light travels in a year. The speed of light is about 186,000 miles per second (about 300,000 kilometers per second). That means a light-year is about 5.9 trillion miles, or around 9.5 trillion kilometers.

So how big is the Milky Way galaxy? About 100,000 light-years across. That's 590,000,000,000,000,000 miles!

Our solar system is about 27,000 light-years away from the center of the galaxy. The Milky Way is only one of billions of galaxies in the universe. The closest galaxy to the Milky Way is the Canis Major dwarf galaxy, discovered in 2003, which our galaxy is tearing apart! It's actually closer to our solar system than the center of our galaxy is. Some scientists don't think Canis Major dwarf galaxy is a galaxy, just a trail of stars related to the Milky Way.

Scientists use special tools, such as the European Space Agency's Gaia telescope, to measure and calculate the size of the Milky Way.

compare and contrast

The Canis Major dwarf galaxy is about 25,000 light-years from our solar system. Earth is about 27,000 light-years from the center of the Milky Way galaxy. Is Earth closer to the center of the Milky Way or Canis Major?

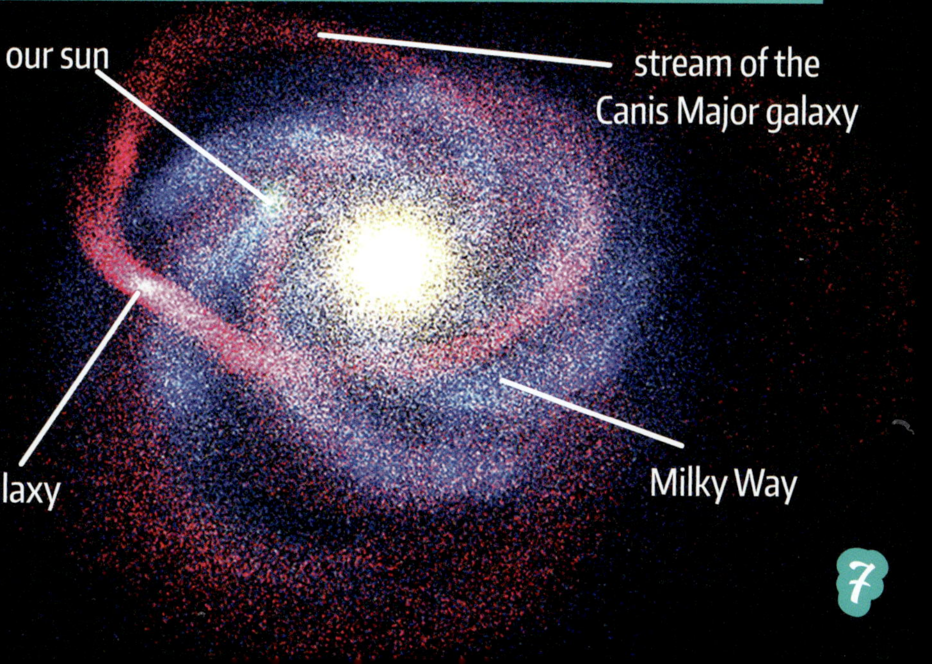

Scientists call the Canis Major dwarf galaxy an irregular galaxy because it doesn't have a usual shape.

Spiral Galaxy

Galaxies come in many different shapes. As noted earlier, the Milky Way galaxy is shaped like a pinwheel. It's called a spiral galaxy. We can't see that from Earth. Astronomers have figured out the shape of the galaxy by mapping the movement of stars and hydrogen clouds within the galaxy. The Milky Way has two major spiral arms and two minor arms.

Just as the planets in our solar system revolve around the sun, everything in the Milky Way rotates around the galaxy's nucleus. It takes about 225 million years for the sun to make one complete revolution. Other galaxy shapes include elliptical, lenticular, and irregular, among others.

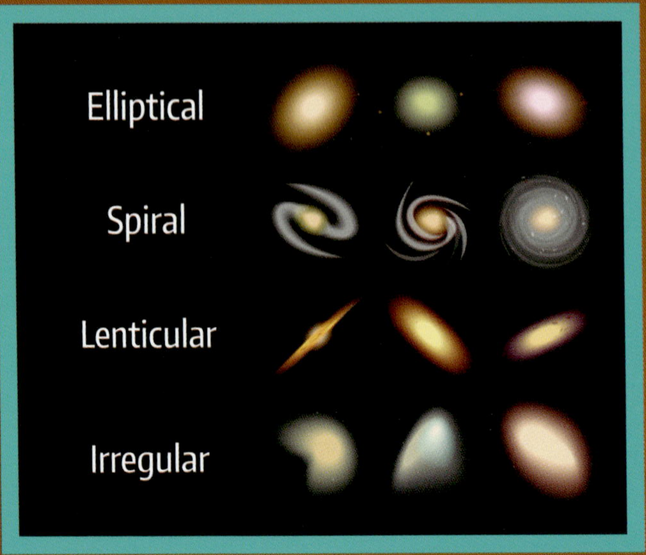

There is variety even within the different types of galaxy shapes.

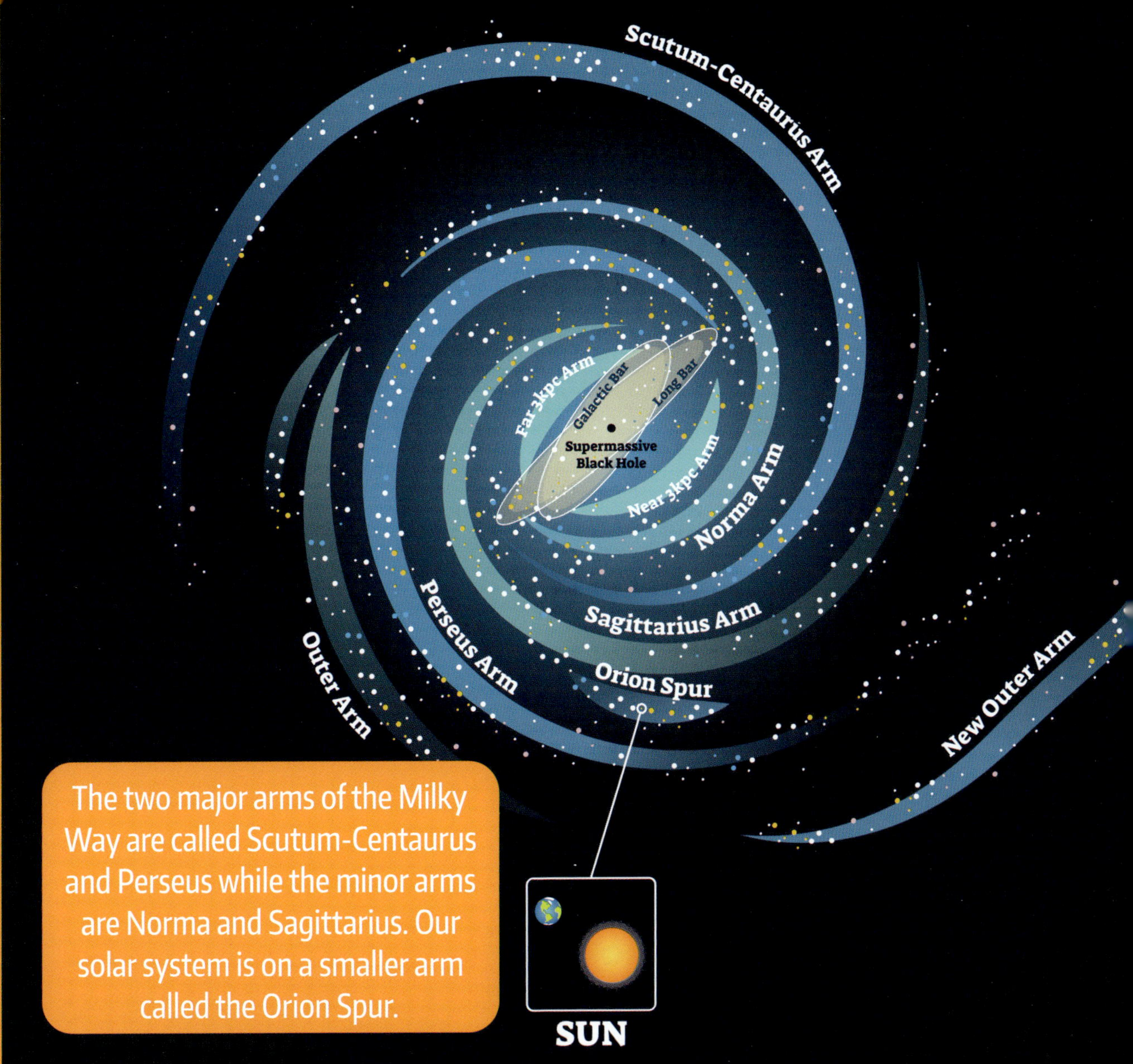

The two major arms of the Milky Way are called Scutum-Centaurus and Perseus while the minor arms are Norma and Sagittarius. Our solar system is on a smaller arm called the Orion Spur.

compare and contrast

Not all galaxies are spiral like ours. Look at some photographs of galaxies and compare them to the Milky Way. What other shapes do you see?

How Old Is the Milky Way?

It's hard to say exactly how old the Milky Way is. The oldest stars in the galaxy are about 13.6 billion years old. This is close to the age of the entire universe, which is estimated at 13.8 billion years.

But even after the oldest stars in the Milky Way were born, they still had to be drawn together into clusters, or groups, that in turn joined together to create the galaxy.

The bottom picture shows what an arm of the Milky Way would've looked like from a planet about 10 billion years ago. There weren't as many stars then.

While it was forming, the Milky Way galaxy crashed into and absorbed, or took in, at least two other galaxies. It's expected to crash again, into the Andromeda galaxy, as seen in this picture.

As a result, the Milky Way is a bit younger than the oldest stars in it. In 2022, astronomers realized the disk of gas, dust, and stars that surrounds the Milky Way's bar-shaped center region was older than they used to think. It began forming about 13 billion years ago.

Consider This

The Pleiades is a star cluster. These extremely bright stars formed within the last 100 million years.

Full of Stars

From a distance, many galaxies just look like blobs of clouds or dust. But galaxies are mostly made up of stars. A star is a huge, glowing ball of gas, usually hydrogen. A star is so large that its strong gravity changes the hydrogen at the star's center into helium. This process, called fusion, creates a huge amount of energy that causes the star to shine.

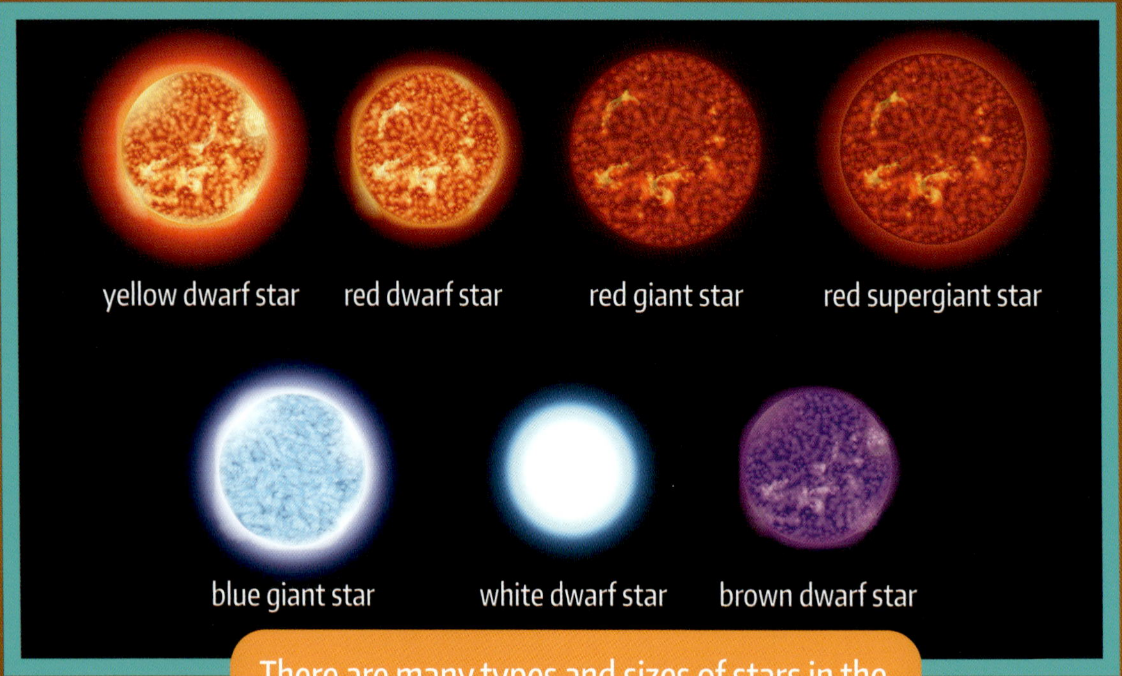

There are many types and sizes of stars in the Milky Way. Our sun is a yellow dwarf.

The Milky Way forms new stars in areas called molecular clouds, or star nurseries. This is a telescopic image of the nearest star nursery to Earth, the Taurus molecular cloud.

Stars are being lost to large explosions called supernovae or falling into themselves to create white dwarfs, but new stars are also being born. The Milky Way grows by about seven stars a year. Scientists estimate the Milky Way has changed about 90% of its gas into stars so far.

compare and contrast

There are between 200 and 400 billion stars in the Milky Way galaxy. That's a lot, but as galaxies go, it's nothing special. One of the largest known galaxies, IC 1101, has more than 100 trillion stars! How does the Milky Way compare in number of stars to other galaxies?

Cosmic Dust

Stars aren't the only thing in a galaxy, of course. Another major part of the Milky Way is cosmic dust. This dust is so thick between us and the center of the galaxy that we can only see objects up to about 2,000 light-years away. That's only about one-fifteenth of the way to the nucleus.

For the same reason, we can only see the dust clouds closest to us. They can be a million times the mass of the sun and can be as large as hundreds of light-years across.

A dark band of dust called the Great Rift blocks part of our view of the center of the Milky Way.

Cosmic dust is responsible for a glow you can sometimes see in a very dark sky. Light from the sun reflects off this dust creating something called zodiacal light or a false dawn.

Some of the pale glow that we see when we look at the Milky Way is starlight bouncing off these clouds of dust. The dust within a galaxy is called interstellar dust. It's usually made up of gaseous elements that come off newly formed stars and change into small bits as they cool.

Consider This

The Latin name for the Milky Way is *Via Lactea*. That means "Road of Milk." Why do you think it's called that?

Cosmic Gas

In addition to stars and dust, the Milky Way has a lot of gas. Gas and dust together are thought to make up between 10 and 15 percent of the normal matter in the galaxy. The gas is spread very thin among the stars, filling the space between them.

The gas consists mostly of hydrogen. There's a lot of it because the galaxy is so big, but it's very thinly spread. In fact, the average density of gas in the universe is only about 1 hydrogen atom per 106 cubic feet (3 cubic meters). That's a density about 27 times less than the air we breathe. The gas between stars in the Milky Way is denser than the average across the universe, but still nowhere close to the air on Earth.

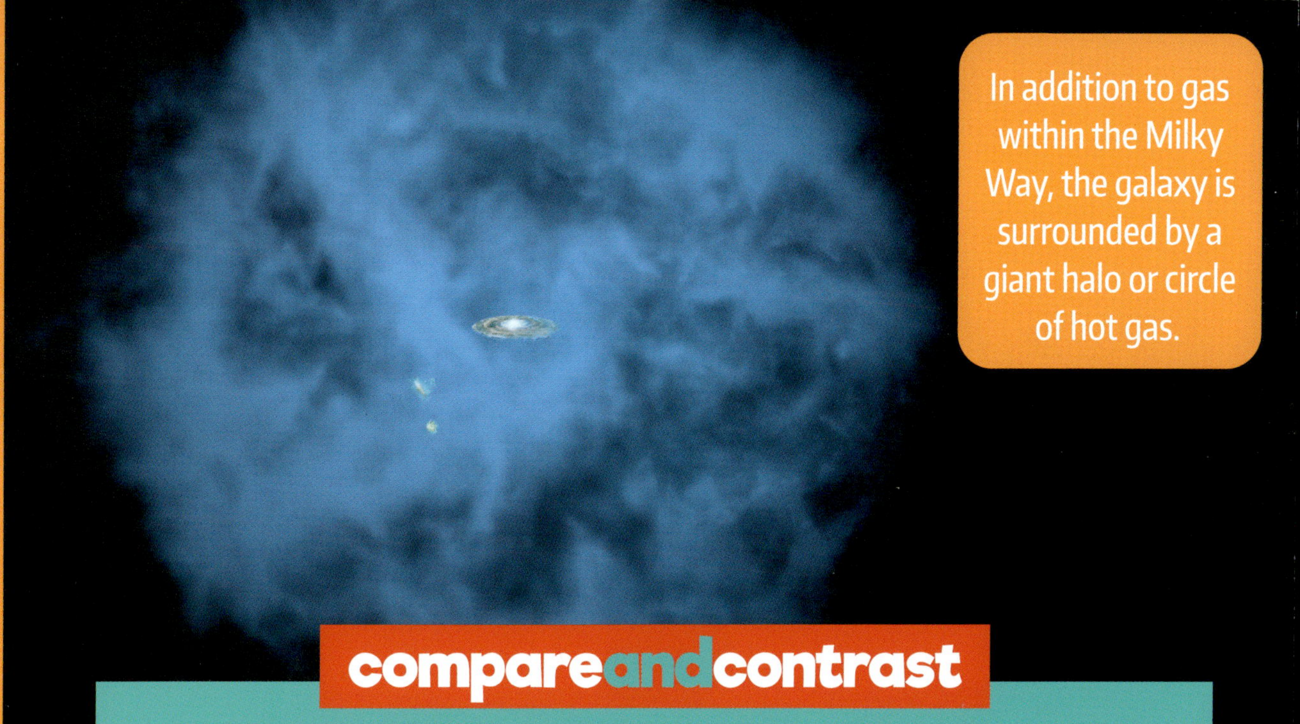

In addition to gas within the Milky Way, the galaxy is surrounded by a giant halo or circle of hot gas.

compare and contrast

Just how thin is interstellar hydrogen? Compare it to ordinary air, in which there are 1×10^{19} air molecules per cubic inch. That's a 1 with 19 zeroes after it!

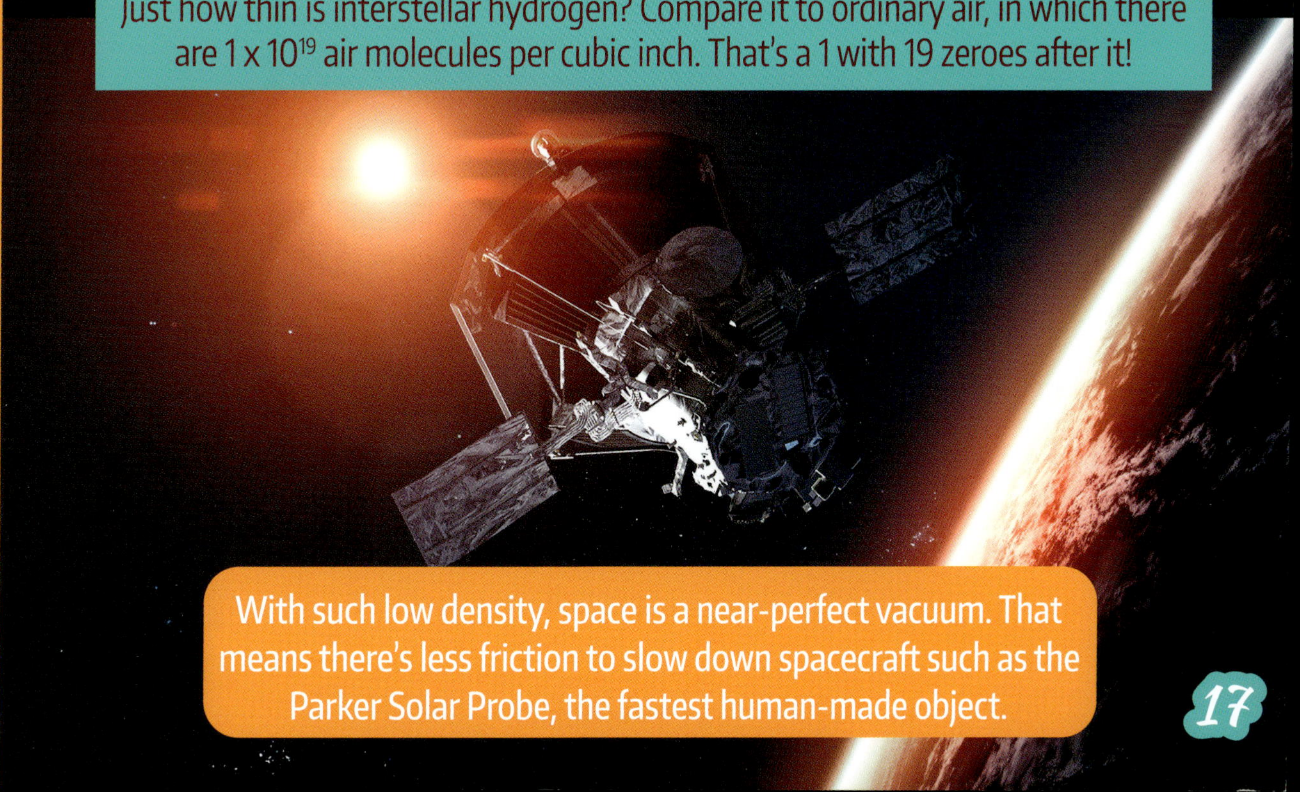

With such low density, space is a near-perfect vacuum. That means there's less friction to slow down spacecraft such as the Parker Solar Probe, the fastest human-made object.

Planets and Exoplanets

Some of the extra dust and gas left over from the creation of stars goes on to create planets. Our sun is a star in the Milky Way that is orbited or circled by several planets, including Earth. Many of the other stars in the Milky Way are also orbited by planets called exoplanets. Scientists are not sure how many because they have only been able to see planets orbiting other stars for a few decades. However, it appears the number is in the hundreds of billions.

There are eight planets in our solar system, along with at least five dwarf planets, including Pluto.

With up to 400 billion stars in the Milky Way, imagine how many exoplanets, such as Kepler-452b, there must be!

Many of those planets are much larger than Earth (like Jupiter) or much colder than Earth (like Mars) or much hotter than Earth (like Venus). But one recent guess is that there could be as many as 40 billion Earth-sized planets that are about the same temperature as Earth. This means that they could possibly support life! Scientists say an exoplanet called Kepler-452b is the most Earth-like exoplanet they've found.

Consider This

Earth is the only planet known to have life on it. Do you think it's possible that life exists on other planets?

Black Hole

At the center of the Milky Way is a supermassive, or really large, **black hole**. Most galaxies have a black hole in their nucleus. The black hole at our galaxy's center is named Sagittarius A*. Its mass is about 4.3 million times greater than the sun's, and it is about 27 million miles (44 million km) across.

This is a drawing of what the black hole at the center of the Milky Way likely looks like.

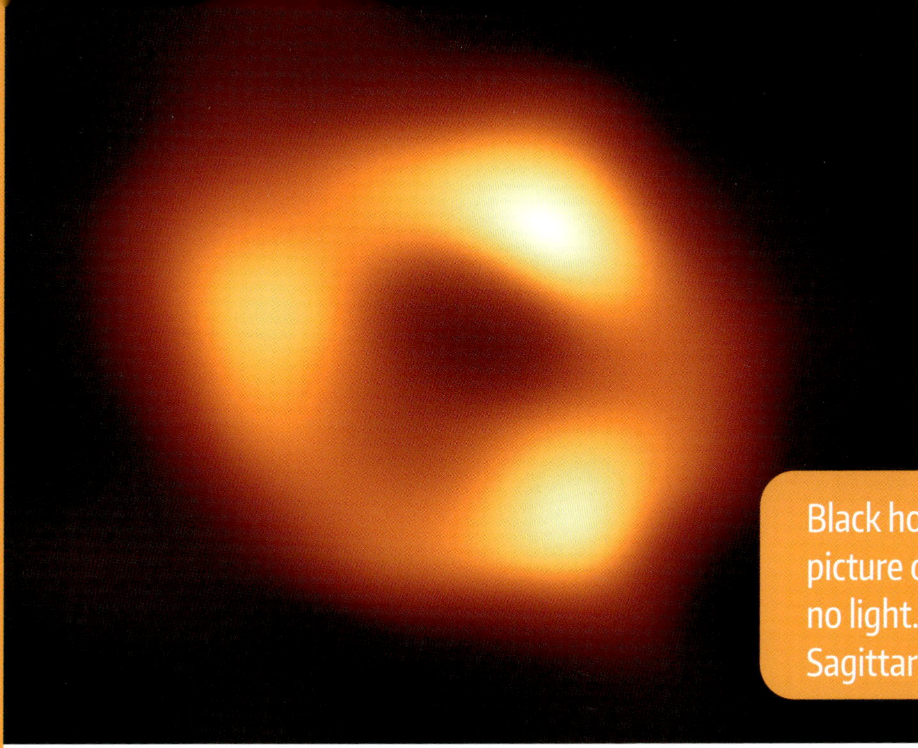

Black holes are hard to take a picture of because they produce no light. This is the first image of Sagittarius A*, taken in 2024.

We can't see the black hole, but we know it is there. We know it because of fast-moving gas clouds in the area and because of the infrared radiation being let off when gas and dust fall into it. As gas nears the black hole, it is squeezed into a fast, rotating disk. This spreads out somewhere between 5 and 30 light-years from the black hole. Scientists believe the mass of a black hole is so large it can actually slow time!

WORD WISE

A BLACK HOLE IS A BODY WHOSE GRAVITY IS SO STRONG THAT NOTHING CAN ESCAPE IT—NOT EVEN LIGHT.

The Galactic Bulge

The Milky Way isn't just a flat disk. The main disk of the Milky Way is about 1,000 light-years thick. However, at the center of the galaxy is a peanut-shaped mass of stars that bulges out above and below the disk. Most of the stars in this mass are very old.

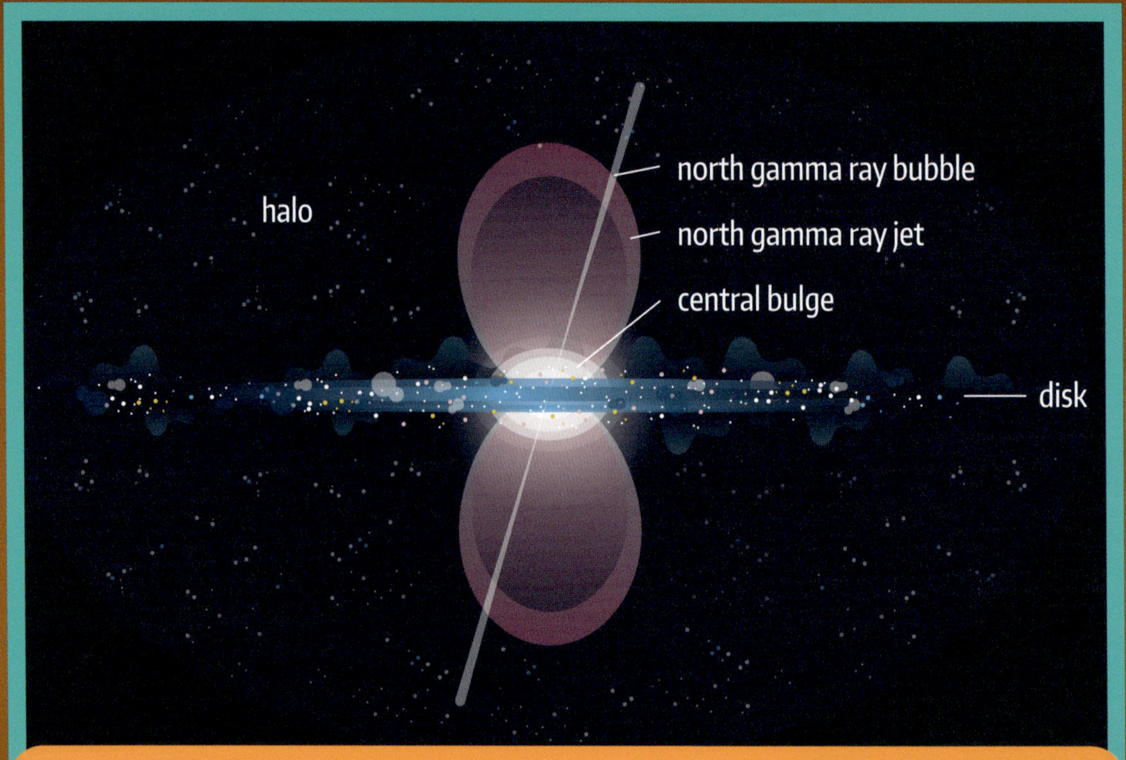

In addition to the bulge of stars at the center of the Milky Way, a peanut-shaped "bubble" of gamma rays, a type of energy, comes from the nucleus.

This is a Hubble Space Telescope image of the stars in the Milky Way's bulge.

Although we can't see the center of the galaxy from Earth because of clouds of dust, we can see some of the stars in the central bulge because they stick up above and down below the dust.

The space above and below the disk of the galaxy has a thin population of old stars like those in the central bulge. Younger stars in the bulge orbit the nucleus in a faster and more regular manner than the old ones.

Consider This

What would it be like to live on a planet orbiting a star far above or below the disk of the galaxy? What would you see when you looked up at the night sky?

Star Clusters

Unlike many other stars in the Milky Way, the sun is a single star. It's not connected to any other stars by gravity. Other stars, though, are **double stars** or part of a group called a cluster. There are many groups or clusters containing anywhere from tens to thousands of stars.

The Milky Way contains more than 150 ball-shaped groups of very old stars called globular clusters. They range in size from 10 to 300 light-years across.

This is the densest known star cluster in the Milky Way, called the Arches Cluster.

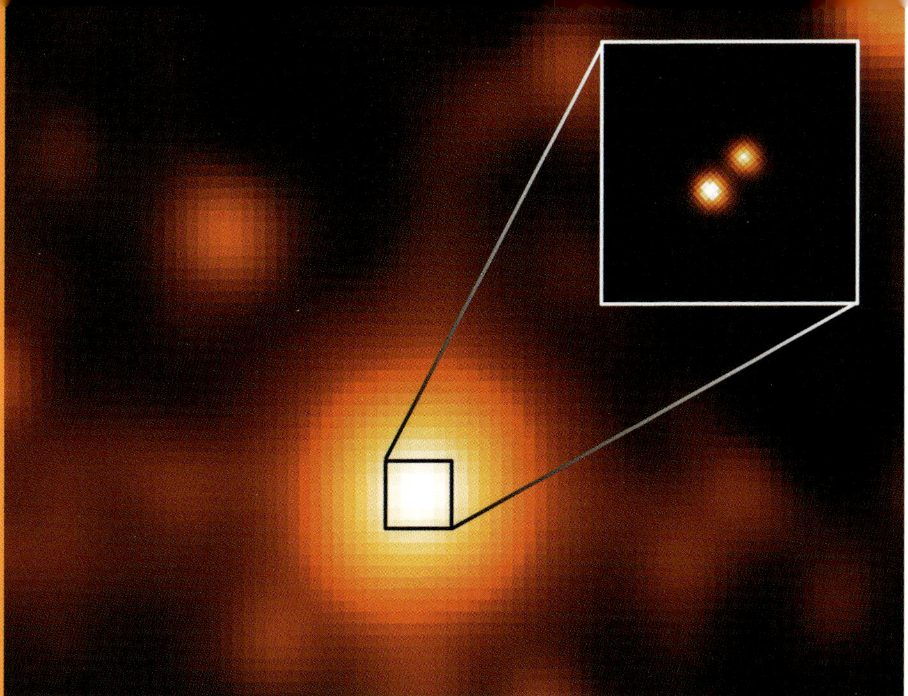

It can be hard to see a double star pair from Earth since the stars can be close together. Scientists first thought this was a single star, but a better photo showed it was actually two.

Open clusters are less tightly organized. Most are less than 5 light-years across. They are much younger than globular clusters.

Stellar associations are very loose groupings of young stars that all formed at about the same time. They average about 250 light-years across. There are also embedded clusters, which are young stars grouped together with a cloud of dust and gas around them.

WORD WISE

DOUBLE, OR BINARY, STARS ARE PAIRS OF STARS BOUND TO EACH OTHER BY GRAVITY. ABOUT HALF THE STARS IN THE MILKY WAY ARE DOUBLE STARS.

The Galactic Halo

Galaxies are even bigger than what we see in pictures of them. The Milky Way galaxy contains a lot of matter that we can see, but it contains even more that we can't see. This is called dark matter (because we can't see it), and nobody knows for sure what it is. We only know it exists because of the effect that it has on the galaxy's rotation.

This side view of the Milky Way shows the central bulge and a picture of the stellar halo around the galaxy. Darker blue areas have fewer stars than the lighter blue areas.

NASA is currently using the Fermi Gamma-ray Space Telescope to study dark matter and find out exactly what it is.

This dark matter forms a large halo around the galaxy, but we don't know its exact shape or how far it extends. All we know is that it stretches beyond 100,000 light-years from the center, and its mass is several times greater than all the rest of the galaxy! The Milky Way also has a visible halo lightly populated with stars.

Consider This

The nature of dark matter is one of the major questions modern astronomers are working to answer. What do you think it might be?

Our Home

When you think about it, it's amazing what humans have learned about the Milky Way in just the past several decades. The next time you look up at that milky glow stretching across the night sky, think about what you're seeing: thousands of stars stretching between you and the center of the Milky Way galaxy, and clouds of dust lit from behind by the billions of stars that you can't see.

Here you can see the Great Rift dust cloud blocking part of the Milky Way from Earth's view.

Scientists keep making more advanced telescopes that let them learn more about space.

It can make you feel very small when you realize that even though the Milky Way galaxy contains hundreds of billions of stars and hundreds of billions of planets, it is only one of billions of galaxies in the universe.

But it's the most important galaxy to us because it contains our sun, our solar system, and our Earth.

For us, the Milky Way is home. And as we look and travel farther into the universe, the more we'll learn about our home.

Consider This

Why do you think it's important for us to study the Milky Way and the universe in general?

Glossary

astronomer A scientist who studies space.

density How compact something is or how much mass it has.

friction The rubbing together of matter that reduces motion.

gravity A force of attraction between particles or bodies that have mass.

helium A colorless gas that is the second most abundant and second lightest element in the universe after hydrogen.

hydrogen A colorless gas that is the most abundant and lightest element in the universe.

infrared radiation A form of electromagnetic radiation that lies just outside the visible spectrum of light at the red end.

molecule The smallest possible amount of something that has all the characteristics of that thing.

nucleus The central point, group, or mass of something, such as a galaxy.

planets Heavenly bodies other than comets, asteroids, or satellites that travel in orbit a star.

revolution The time it takes one object to complete an orbit around another object.

rotate To move or turn in a circle.

solar system The sun, its eight major planets, the dwarf planets and small bodies, and interplanetary dust and gas under the sun's gravitational control.

telescope A device that lets you see far away.

vacuum A space completely absent of any matter.

For More Information

Books

Cock-Starkey, Claire. *Lore of the Stars: Folklore and Wisdom from the Skies Above*. Beverly, MA: Wide Eyed Editions, 2023.

Collins, Ailynn. *Investigating the Milky Way and Other Galaxies with Velma*. North Mankato, MN: Capstone Press, 2023.

London, Martha. *Explore the Milky Way Galaxy*. Minneapolis, MN: Kids Core, 2022.

Websites

NASA: Galaxy Basics

www.science.nasa.gov/universe/galaxies

Read more about the Milky Way and other galaxies on the official NASA website.

National Geographic Kids: Milky Way Galaxy

www.kids.nationalgeographic.com/space/article/milky-way

Learn more about the Milky Way and watch fun videos about outer space.

Index